Country Advanced Defence Program

Kevin Kondol

Published by New Generation Publishing in 2023

Copyright © Kevin Kondol 2023

First Edition

The author asserts the moral right under the Copyright, Designs and Patents Act 1988 to be identified as the author of this work.

All Rights reserved. No part of this publication may be reproduced, stored in a retrieval system or transmitted, in any form or by any means without the prior consent of the author, nor be otherwise circulated in any form of binding or cover other than that which it is published and without a similar condition being imposed on the subsequent purchaser.

ISBN
Paperback	978-1-80369-934-9
Hardback	978-1-80369-935-6

www.newgeneration-publishing.com

New Generation Publishing

1992

Still in this world is need for better country defence becouse of yet not so advanced society counting in this moral stage development. To blame all on system is not proper becouse system itself will not change anything if people themself dont change. That is why in present is better to be prepare until better solution will come than war. For this somehow look is long way yet. To develop better way of defence that is already will make country more secure in case of conflict when such preperation will develop weapons less desctructive but very efficient, will mean better for world.

So if want make country at least one of the strongest in the world can be to it actually use few means technological and strategical value. For example could be drop viruses and phytophags over industrial plants and if such crops production\country will be affected by this it can do much damage to economy.

Also with this stage of technology in all world of computers can be apply so called computer viruses and there is really very large number known such viruses which when apply in chosen country or countries can make lot of damage to industry at all. One of the way to introduce is by weather computers or other internationally use one means. Also can be use as a weapons blood infected with AIDS or other more dangerous viruses especially on chosen area of millitary points to weaken enemy army. Simply way is too use just needle on touch body so by this way many can be infected. Beside this could be blow up some smaller nuclear power stations which can bring much more efecttive result than not so always dirrect yet not always precise bombing. This could be done in enemy country and is better blow few such smaller power stations at ones as later could be make security for other ones. So in short when come to conflict with enemy country what can be done best is to drop viruses early over crops to destroy enemy country more economically and then spread dangerous viruses and bacteriums between mostly millitary targets to more yet weaken enemy country and yet in later stage blow up few smaller nuclear power stations in couple different location and this is final of business, much cheaper and more effective than nuclear weapons and yet if done more properly with secure preperation first- nobody may find who done it in such first stage. Is still better than use nuclear weapons as may do less damge and such strategy is possible to use against much stronger power.

When using laser beams with different frequency one can direct it to sleeping or active vulcano to evenually make them blow up.

Is posible to make present drugs useless against some bacteriums or viruses by for example often repeated use on them as they get inmunise..Then deadly viruses is possible to use in enemy country before some drugs develop agaisnt them yet if developed early it have to be send to different location in own country.

Some radioactive bomb hidden in suitcases can be smuggle to country and set in few places there and when come to conflict few of them can be blow up and then announce that farther few of them will also blow up if attack against own country will continue. And enemy would dont know how many actually will be there located in his territory. This one if done properly small country for example can win against strronger countreis. To make more weak enemy country could be also done with help other countries which produce or smuggle drugs to some western countries and help faster spread them there so people will get more sick and more degenerated, especially if done on millitary so would be not so active in action.

In some public or government places can be set listening devices, tapes to hear important for own country conversation or to get information from millitary computers of enemy so if done in secret may take long time to discover it.

Some of these things can be done with help of third

country or from their territory for better result. In these time atomic weapons may become less important against new weapons, technology and strategy if develop as part millitary program in own country as these small samples show and yet more developed later.

One of the ways to make neutron like bomb is to use radioactivity. During nuclear plant process or accelerators activity is lelft lots of radioactive material which can be later with gas transfer to prepared bombs and missiles with different power of destruction.. Is obvious that more radioactivity will get out after few hours than afet few minutes so to make stronger radioactive bombs have to be make more radioactive clouds by few hours to bombs and missiles then close and use when will be necessary.To make it more stronger and more effective would be use stronger magnetic gas so much more radioactive gas mixed together with magnetic gas will accumulate inside bomb which walls have to be yet more magnetic to attract less magnetic radioactive gas. During bomb explosion these magnetic properties would be lost, especially with time of keeping them inside.

But still will spread around. Also instead can be use pressure to accomodate more such material inside bomb. Such radioactive bombs and missiles can be very small but spread around from few hundred meters in diameter to much more if need.

Also is know how much pollution can cause break in nuclear power station so is no doubt how much devastation can do radioactive bomb which after explosion will release large amount radioactive gas around. So simply have to be produce radioactivity even in nuclear power station then these radioactive substances have to be transfer by few minutes or few hours to prepare already bombs. Process in power station is already known so is easy to do it then such bombs have to be close and then ready to use if will need for it. During explosion neutron bomb mostly is devasted biological organisms without matterial devastation , but is much cheaper to make radioactive bombs and dont need much technology only material which can be done in laboratory and then to send few o bombs, tens small rackets with such radioactive bombs to enemy country and after explosion most biological organisms could be destroy but still better and safer than neutron bomb.

One dont need atomic as these one will provide similar effect, especially that most people die not from bomb explosion just radioactivity.Though have to be carefull not to release to atmosphere.to much.Such radioactive bombs can be made from radioactive material which have half of life span so later after fall down it can be populated in time by other people. If for example enemy would have over one million soldiers then if will be use 19 or more such radioactive bombs or missiles so whole army would gone this or other way in minutes or hours and war could be over. Such radioactive gas is very light then when concentrate it can make small but dangerous bombs which can be use even from ships to attack enemy country on on ships. Put them also outside territorial waters such coutries on bosttom of sea well hidden before expected conflict. Some may put them also in satellites and send from space over attacted country even if not suppose to be done according to treaty. Even if some of them will be shut down radioactivity will fall over enemy territory. Development these bombs will make country kind of nuclear superpower but for much less cost and technology as when use over country or army they will do the same destruction or similar like atoms bombs and can be use against larger numbers of planes with explosion such missiles.

Better and save more way of defence to apply on any attacking power is simple use of blood infected with AIDS or other more or less danderous virus.

Especially well train agents can be send to attacking country and then put these dangerous long lasting small amount in blod mixed viruses in factories in in boilers to popular drinks so it will be infected. Or for larger scale effect to maajor water supplies in such coutry to pipes where they are already pass save test that is underground with needle size hole to inject there. Of course is better do it in places where are millitary personel or factories connected with millitary production all to slow down later millitary action. This way done by one or few agents carefully so in relatively short time large part population in these areas will be infected and by this weaken millitary power. It is safe more for country which is farther distance from attacking country. Of course is not best way but as about every weapon kill so such one may even limit this in dependence how is done, yet is much cheaper and better than bombs- especially it may help smaller country against stronger one if need come. This second way is safer, better than first cost yet much less and part country can be kind destroy and some may not know who even done it. In this century as come have to be use weapons and strategy different if country dont want be left behind in this kind weapons too. Will be not need for large weapons. These one do better and surprise is important part beside.

Is of course better never use it but so far as development go it show that dont yet advance very far in many aspects so always better to be prepare.

Also is good opportunity to get nuclear and other scientists from other country and pay them very vell which is still profitable especially if they get results.

One of the way to improve energy output in atomic and nuclear bombs is by try to excite atoms to produce more cncrgy., though better result are achieve by application antiparticles – like anti hellium, anti hydrogen becouse with these application can be achieved enormous amount of energy. These antiparticles is possible to made in laboratory in smaller or greater degree. Beside could be applied process which happen in some type of stars when they going to die. Their external parts kind of blow or lift outside becouse nuclear forces inside get more weak and could not hold longer outside part as core of star is to weak for this. Somhow it can be done in laboratory on much smaller scale. Yet such process can be make more faster including by concentration and change of temperature and with different speed of spin around. To transport nuclear weapons to distant countries is not such problem now. Most rackets can do it if they have enough fuel but some elements like uranium can make enough fuel and is some way to improve it.. Also can be done with fast super sonic jets over long distances about as fast as long time ago intercontinental missile. And yet to make sure they will get there such super fast jet could be equiped with few antimissiles so when over enemy country territory other missiles would not destroy jet just to be destroy by own in jet missiles. Do not mention that these missiles is possible to send from ships as well.

With present development laser can shut up to few kilometrs to destroy target so they could be place in planes or rackets and when missiles or intercontinental missiles come close to planes or own country they can be in short time destroy even if shut from ground as well. Also very few fast small missiles can be use with very strong explosion power and hit the same time intercontinental missiles or planes, yet these small missiles with very high explosion power can be also send from satellites if necessary in lover or higher latitude from over own country, or even from weather balloons which are often fly high too.

Important however that they will hit such big target the same time to have desire effect.

These often make terrorist acts like bombing and similar are very uneffective and basically they are waste of time. There are much more effective ways to destroy enemy when come to conflict. They are known from time to time breaks in chemical plants and factories which sometime by longer period release some chemical components over city or cities and this can be notice also by smell of these chemicals over large territory or big city. Similar way could be made not smelly and not dangerous gas in very denset amount and put them to already made bombs by pressure or in solid state to get most amount such gas and then heat when inside bomb turn back to gas state which culd be very large amount of it. Then few such bombs can be place in some points of millitary or part of city where they produce for millitary, and then even without explosion release to air over territory. Such average volatile gas will not harm anybody not even have to be notice, but can be mixed with other more or less volatile gas or substances made to gas. To make these gases well mixed and spread together they can be slightly magnetise or other kind attraction between them can be make so they will spread evenly.

To such volatile spreading gas could be add and mixed in mentioned way other poisonous, cancerrigenic substances or substances which cause permanent or temorary blindness or some brain damage,even temporary, or these one which cause infertility among people. Also can be spread viruses, bacteriums- first by place them in some chemical frame, then slightly magnetise this chemical frame and mixed very well with also with slightly magnetise volatile gas to get them stick together.This way lot viruses and bacteriums would be spread over designated area with help of volatile gases and nobody may even notice it. These viruses not have to be deadly or just do some damage to organism or just case infertility between people without killing. Have to be make sure that enough both gases will be inside prepared bombs to spread over area smaller or bigger. They can be smuggle by ships or planes if not from own then from third country and there place in desired points and eventually release to air if time come for it. For larger scale if this is well prepare can destroy whole country or badly damage it in dependence also from applied substances or viruses yet here is choice by thousands so is no point explaining them as they are already known.just can be mention that up to it can be add volatile, hightly inflamable gas to cause lot of fire especially in dry weather. Even as is sometime done, that is bombing itself- this can be much more effective. It is no point to bomb some areas yet populated as this not always bring desired effect just damage.

Instead for example could be made few strong bombs to enemy country and in few places there where are power stations and blow them out all about the same time for safety reason in few points enemy country . It may dont kill almost nobody but if properly done such way it will take long time to rebuild them.

It will couse much damage other way, distraction and cost lot of money for enemy country leave whole cities or region without electricity by at least sometime few weaks, again in dependence from scale of destrucion make yet lot of chaos.This will do much more damage than all terrorist acts together. In war time country can do that with help some extremist groups from other countries and put blame yet on somebody else.

In one experiment to make missiles fly farther was use long tube in end of misile, rackets where gas come out after burning fuel. In the end was place blades which spin around becouse of pressure by coming gas to the end of tube .These spinning blades turn small wheel which turn extra engine in missile, racket and by this was made extra power to move missile or racket even with help electricity. So simply was use presuurec made by gases coming from inside after burning fuel so it done extra work to provide more power to fly yet farther. It could be done few ways and dont have to be put tube in the end flying vehicle to turn blades made with hard material for temperature resistance, but main goal here is to use power pressure of gases from racket, engines which other way is wasted. It have to be done carefull without effecting possiblity some pressure back to racket engine, so gas have still move freely to space turn only blades, wheel or providing power, pressure which can be use as extra source to move missile. There are few extra technical problems but they can be solve by known technology. This way is possible to move missile at least for much longer distance in dependence how is done and faster by use power of wasted burning fuel or gas. Beside racket can use solar power to move missile or hydrogen power from water or combination all these.

To the fuel can be add radioactive or poisonous gas so when reaching enemy teritory radioactive or poisonous gas can be release together to air with standard gas coming from burning fuel to make extra damage to enemy millitary. Lately was done experiment with narcotics drugs which some part of population use There where are going sources of them as crops was done different genetic material which could do more brain and body damage. So if on large scale in these countries where these plants grow for such drugs will be introduce new breeded plants it can damage more part of population connected with millitary with very cancerogenic, brain damge. With such wide habit drug use it will have large effect on such part population.

Before any conflict own country have to be prepare some ways like use of water as source energy to extract hydrogen from it and apply in power stations or industry as is already ;little done here and there saving on import of energy source.

These new energy and clean technology is not use becouse of business and yet could be improve and develop to own need. To use more improved grain crops new methods and technology is possible make more sufficient own country in food for long time instead import it.

None of conventional weapons can beat well organise strategy of use viruses.

Beside mention before can be add that some samples dry blood with more or less dangerous viruses may in small amount add to some narcotics which then smuggle to enemy country. Some agents can be employ in place where their production is going then pass to design destination country which attack. To drugs which are going there to be use by injection can be add some not contegous viruses and to drugs which will be smoke can be add some very cancerogenic that is cancer causing substances even little radioactive.Though importand is not to overdue or give to high doses becouse sooner can be discovered. By such spread of using drugs it will be very easy to destroy large part designated population and make more weak country with target of with millitary connection, these one, yet to try avoid civilian population the same time. Not contegious viruses are not treat to spread to other countries yet when enemy country is more seperated from others.

If atoms or other nuclear weapons will be send and use it will be know who send them but here may not be even know who done it and even if it take longer in this case it is more cheaper and safer to use it and will do not do as much damage like nuclear weapons if more safe way done. In some of time will come out cure for them.

As to other weapon is possible in change of atmospheric presure for example to make stronger and even faster moving cyclones, hurricanes. Some may have been try to use before in some areas.

Also can be mention that when necessary under sea telephone cables between countries or continents can be cut in one or few places for greater damage so to cut some of communication . And some small bombs with very volatile fast spreading inflamable gas can be hide in enemy place close to millitary, then release and when sufficently spread out add radioactive gas if needed.

During attack by enemy on own country can be before spread instead mine field radioactive field almost not penetrable to land army as if is spread over wide strap of land will be dificult to overcome becouse of radioactivity everywhere.

Long one of course and wide enough or just under surface place radioactive sheet which can belater remove.

Place there suitcase bombs yet radioactive in few important targets then if more need blow extra some nuclear smaller power station. Also by some missiles and planes drop some radioactive bombs over enemy army and ships. Yet then after announce that there are still more hidden bombs in country though could be only few of them but no one may known and say if atack will continue they will be exploded. If need for this repeat action again. With so large public outcray would be doubtfull if attack will foloow. It is saving on money, own people and with good preparation can beat strongest army by thestrroy dierect not only far away army as it be not enough. Even small country can do this and in similar fashion against strongest one.

It only mention in general therms but with more details and better preparation plus these mention before other weapons and strategy it is no any doubt how effective it can be. Is better not use much or at all in small conflict instead better get ready to use them when really will be necessary even against strongest enemy. It is not question of morals when come to defend and beside it is much better than nuclear weapons which would do much more destrucion.

To make more controlled these explosion in laboratory magnetic radiation could be also use. When average explosion happen is often rapid and not contoled, but such gas can be magnetise so later can be put in strong magnetic field. If this explosion happen strong magnetic field will somehow hold this rapid explosion and can be send the same time strong magnetic radiation or magnetic explosion on much control level as opposite physical force which would be even directed in desired way. By such way it will make more nuclear explosion or rather thermonuclear which is need to create very high pressure and temperature to overcome electrical repulsion more close to nucleus particles.Such magnetic radiation is possible done on subatomic level. Stronger or more controlled magnetic radiation or explosion with different vibration and vaves lenght as opposite to termonuclear explosion will be achieve. This way could be prepare heavy hydrogen and put inside thin ball which will be explosion material and beside nucleus or whole hydrogen can be magnetise the same like explosion material, then put in to next similar device what could be repeated few time with last ball at least magnetise or with ferromagnetic material so during explosion it will attract more inside creating more pressure toward inside.

Is possible to create repeated wery small explosion so radioactivity will comodate inside ball, vessel. Also of course can be use few units of balls, vessels connect and exploded together same time to create much bigger yet explosion and radiation spreading around.

Yet more simple way to create radiactive bomb is to transfer radioactivity from nuclear power station or to take radioactive core and by heating turn it to gas then put it directly to bomb. Such core or radioactive material could be radiated magnetically so if later in bomb could be make rotating magnet more then radioactive magnetise or ferromagnetic matterial will get inside bomb.

Also if then happen failure in nuclear power station then if such some magnetise radiation will get some out can be hold to great degree by thin magnetic sheets which will atract magnetic radiation becouse some stronger magnetisation, so it will not spread around out of nuclear power station.

Beside if such radioactive magnetise bomb if use and explode then when will spread around will be more easy to get out of ground by stronger magnetise thin sheets spread over ground . Also any radioactive material is possible heat and turn to gas and put to bomb by similar way of magnetisation or yet by use of pressure to accomodate more inside bomb.

One of somple known process of makig nuclear explosion is by put some nuclear bomb or heavy hydrogen in ball, vessel which is also typical explosive matterial. After exlosion will follow implosion with higher temperature so electrical barrier of repulsion is overcome and thermonuclear bomb is start to glue itself. Becouse of it energy start flow and explosion follow.

Such simple nuclear weapon dont have very high power of destruction but process of bigger explosion could be improve in few ways. One more simple way is to put heavy hydrogen in one ball, vessel then start to put similar in other ball, vessel and then in next one or few similar balls with biggest one in middle or center, so when first inside ball will explode it will blow next vessel with heavy hydrogen and then next one until last one which will receive strongest power of explosion from former procceding explosions and will have biggest power of explosion itself. Also all ball bomb could be blow about the same time or they could be connected with few other similar units of vessels so when blow together they will have higher power of explosion.

Could be done too in reverse process but important too is that if will be more vacuum space between balls or vessels.

Other way of improvement is to radiate strongly heavy hydrogen or nucleus with magnetic ways so whole lot or just nucleus will be type bomb more magnetic.

Then to surround it around with very strong surface or mirror like which will reflect back properties of explosion like sound waves for instant accustically increase-and surface itself have to be as much as possible magnetically strong so wth explosion will be stronger pressure toward hydrogen. If mirror will be destroy during explosion it will not matter. Particles are more easy to radiate Already have been done experiment in laboratory where by lovering temperature close to absolute zero movement of atoms particles been slow down do not mention about possibility to go below absolute zero, one way by creation of yet bigger vacuuu. These particles are more easy to radiate. By longer radiation also from nucleus is possible to change electrical forces so they will not repulse each other and by magnetic radiation they yet can become more magnetically glued to each other so less energy escape outside bomb during process, yet with help with explosion already describe. Magnetically radiated can be other particles neutron and proton. In atoms if neutron dont escape outside bomb it can help make bigger chain reaction. So when will be place around very denset magnetic field its inside part will act as oposite pole to magnetise neutron particles.

This way with strong magnetic field around neutron will keep inside help to create chain reaction. In neutron is have to be done opposite magnetic field around, it will attract them to get out. The same way is possible to get out more protons to create kind of proton bomb. The stronger magnetic field around the more particles will hold inside or get out in dependence from type desired bomb.

In laboratory is possible to make very small more controled explosions this way, when farther around will be very strong magnetic field or dense magnetised gas.

This way after small explosion radioactivity or neutrons

or protons will accomodate in this strong magnetic gas. Then only by change heavy hydrogen and explosion material is possible to do many explosions until gas become saturated with radioactivity or desired particles. To increase farther spreading such bomb few such gases with magnetic properties could be mixed together and yet add around very volatile gas magnetised with desire strenght the with typical explosion such gas will spread very far around, when shortly will follow explosion radioactive particles which will spread much farther this way by being attracted by magnetise volatile gas. Or they could be exploded the same time. Also beside could be add hightly inflamable volatile gas for farther spreading.

If neutron saturated gas will be use or magnetic field it will be similar in explosion to neutron bomb for instance with much lower cost and time to make it.

More stronger magnetic gas it will hold more particles which first of course will be magnetised too.

It possible yet to use different kind of strategy to do damage to enemy country in case of conflict. One of effective way to destroy enemy industries is dealing with electricity. It happen before when for short time it was black out even for short time but do damage to industry. It can be done similar with good effect by some traiined agents. One of more simple way is to make by sending from power stations everywhere all over where electricity have very high electric current by change it from alternative current. Though it is few ways of doing so it can be make sure that wires by which current flow will be damage, burn out or melted in possible lenght from power station to everywhere and do so even over large part of country.

Such cables or wires can be burn out first little gradually without possible visible effect and then suddenly impose very high temperature to such degree which is known to melt these wires. But is best to make sure that such vires will be burn at least on most lenght and by this to create enormous damage on very high scale stoping basically whole industry most these connect with millitary and it will take long time to repair it all in factories, homes, buildings, everywhere electricity get by cables which later all have to be change. Best is yet to pick hours where most industry and office, transport are working. To do so could be also use laser which with very high temperature would pass along wires which are electromagnetic and with similar laser magnetic properties would pass long way.

On small scale it was already try it on few buildings where was send from outside very high current with yet some extra artifficially add odities that rise suddently so high temperature in all wires that burn them out in all buildings in most their lenght. Beside was try transport some matterial by use production line and push bottom from start to move line when the same time push bottom from other site of transporting line .So it stop becouse of high circuit which burn wires. Similar way can be done on large scale. Though it is most generally mention but such ways to do it are rather known so is only to take few electrical expert to fint most effecient way of doing this and to try first on some smaller scale in own country to find best ways first. It can be mention such strategy as when done properly it can do really powerfull damage to part of country yet these parts connected with millitary with not that much cost when doing it carefully.

Other very good weapon is to use substances, viruses defective genes which make people infertile and use as last action on enemy country when really necessary.

They can be spray in air even from private hire planes in enemy country, yet especially target these places which later could be as dangerous for own country.

Yet they can be introduce to some phags for farther reproduction or just release with average not smelling gas over desiganted place and these phags, defective genes will be breath so eventually with time they will make some groups more infertile so it will make enemy country more weak and less dangerous in time.

They are not dangerous for other countries becouse these viruses, phags are not contegous and it introduce metion way or in popular drinks so this strategy will weaken enemy country without much risk to others.These defective genes, viruses which cause infertility can be done in laboratory then carefully keep until will be need to use them if necessary. Maybe even not becouse may not come to conflict but already many similar or other laboratoy already keep in there.

Also could be mention that is possible from own country to get access to computer in any country in the world. Either millitary or industrial computers to get necessary information. Of course without being catch.

Is good to use some movable computer or even sometime also from ship or else, or send agent to foreign country to get more information. Yet to avoid more or less detection some most developed countries could send part of them to the moon to send signals from there and then transfer to own country.

If own country find out that was infiltrated then could later give more false information. Beside is possible get information without cod,pass word by listen to other working computer it frequency bits and units when in operation in own coutry. This dont kill but can do damage also to country with better development and improvement of them.

Over country territory maybe make small tube with also small engines along and spread there inside line, cord which will be move by these small engines along tube by usually electric power. Then if necessary could be attach to this line any material, packed to be send to any point in country or to other site of country.With good power to move them will be possible to transport very fast any matterial wanted and from safe site nobody could see what was transported unlike transport done on surface, ground. Something similar maybe done in space around earth, though even intercontinental missiles have to travel in space.

Also it could be done by placing line, cord between ships which all together could be even few hundred or few thousand kilometers between first and last ship with some lines between each. On top end of ships with larger motors to rotate wheels very fast. Or small devices can be place just under water in sea and when need rods can be straighten up over water to transport small missiles or else, Similar way as on ground. When missiles reach end of the cord can be fire to enemy country. This especially if own country dont have long range missiles.

Becouse usually war dont happen very fast so few weeks before or few days can be made such preparation which dont cost much and is very simple to do it as all it need is long cord already made and few operation to spread it in desired lenght.. If want to find out deep under sea water other ships or get to bottom of sea, ocean maybe apply big coil with thin but rather bit hard line,cord which could be from coil spread down to botom of sea with last end some over water so will be no pressure then from to. Along maybe attach small cameras inside to oberve on the way to sea bottom what happen around . Beside if want to get deeper under water triangular, conical device better use for lower pressure.

On smaller scale by some individuals have been made with help of stationary balloons long line to tens of kilometers though could be even made few thousand kilometers in lenght, with one end hanging down close to ground . Then in end of line close to the ground been attach small missile and then when wheels have been turn by electric power so on distance tens of kilometers this missile overcome very fast by line attach and strain between balloons. Then could be fire out to target to save on fuel, yet if balloons are on high altitude so when fire from there gravitation will be less than from ground. On scale of country not just by few individuals it can be develop to bigger scale to transport missiles on distance many thousand kilometers to save on fuel- and still with high speed. If faster wheels will rotate then faster speed of missiles will be.

Also can be done on country scale on ground to transport

missiles or something else very fast in case of war. Or can be use also planes which will fly all with the same speed even thousand kilometers apart with line streach between them and transport missiles as fast- as between satellites or ships anywhere to target on earth with still very high speed in dependence from power use. On line between planes over enemy country could be attach also bomb all kind and drop down when distance between planes will be such big that cant be seen planes and not even lines with attached to them bombs.

In space was done trial with mirror which light some area on earth. But then instead could be place radioactive mirror which may reflect radioactivity on designated area on earth. Surface of mirror need to be made with strong radioactive matterial which when put on desired angle with help of optic calculation so it will send radioactivity down. Also can be mixed with on with poisonous surface of mirror which will send both poisonous and also radioactvity over enemy country.so can do damage to biological organims without destruction. Of course have to be done such way that will be not notice by anybody, yet enemy country as well. It is again much cheaper than nuclear weapons doing similar damage to enemy but safer way.

In space it can be done by satellites to place them or on lower altitute is possible attach them to stationary balloons also with mirrors though hide so cant be discover. Stationary balloons on altitute 20 or 30 kilometers still may cover some large area on ground close to enemy territory.

Yet if want to hide something on bottom ocean could be create over any device whirl so by that pressure of watter over will be less or if need below too.

When from missile or racket coming off heat and gas which is wasted so again maybe use blades that will be turn around connected to wheel which by strings will be attach to generator inside missiles or racket and this one in turn will provide electric current to electric motor so by this it will be extra power to move objet farther or faster. Coming of heat and gas have very big power to move very large wheel but becouse cant put there heavy generator inside so by extra gears this power can be use to turn around generator much faster which in turn could be use to produce electricity yet more. Beside that in front of missile, racket maybe made small hole by which strong current of air will pass during object fly and inside then place behind hole blades, wheels which will turn around, connected to the same or other generator. Or maybe yet done on sides with small wings with hole using the same principle and connected to the same or other generator. So all together it will provide much extra power to move object farther ahead. In end of missile even with or without small tube bent up or down could be place small wall geometrically design to special point to return strong gases which fall on it and becouse this strong gas will provide strong power also it could be use with extra equipment to push object much faster without use of electric motor so one power can be transfer to other as it can be made many of innovations and combinations to use this power in many ways for long fly object even many thousand kilometers in dependence from use. So pressure and air flying missile, racket maybe use as very strong source of power including to fire smaller missiles connected sometime to big missile, plus on smaller scale as have been done, use additionally solar power for it. Similar way this could be make to other vehicles as well. Also in outer space rackets - where is not air but there is very strong cosmic radiation. This radiation maybe use similar way by made very sensible blades on which will fall cosmic radiation with yet made in the end very thin wholes so this radiation will be more concentrate when come on blades to spin them around.

Such power extra maybe use not only to move object but beside creating electrical power to use for equipment inside flying object.

With such developed technology now it need only very limited number of people and not much financial spending either to destroy enemy army or part enemy country. Already mention virus strategy is one example if yet more developed.

As with development bacteriums and viruses in own country also have to be develop cure for them which will be of course keep secret in many parts of own country to use if necessary. Then if enemy country will attack so when know about coming conflic early few people can be send there to spread viruses in important millitary places and before when attack begin it just start could be make known to enemy country that only own country have cure for these viruses or bacteriums and if attack will start these cure never will be reveal so large number of people will die though dont have to be true in all case that such number will be affected but enemy country will not know in how many places and on how big scale these viruses, bacteriums have been spread.So is chance by that it will not atack in such situation. Beside can be prepare viruses, bacteriums which dont kill but have many not desirable effect for health. Similar way in addition maybe place few radioactive bombs. So just before attack some of them have to be exploded and of course mention that more will be exploded if attack will start or continue. So if enemy know that their people will be in danger may not contnue with attack. This could be done with combination virus strategy and radioactive bombs with addition strategies mention before which can destroy much industries yet these connected with millitary activity.All this just by send well trained people which will prepare these weapons without any suspision. Though such methods may look to extreme but they are very effective, cheap and dont need very advance technology, when from that known traditional wars are even often yet worst.

Then important thing is not so much to destroy enemy army only which is coming just to hit in heart, in this instance in enemy country territory which send army or attack as these strategies and weapons may do this efficiently. Beside few countries may develop already some of thsese weapons just keep them unknown so own country should be ready and prepared also.

Tactics in war is important too. So is better to change it sometime. For example not just atack few armies of enemy with few armies but left smaller group of army to defend or involve in fight but largest part of army could be send against smaller groups enemy army with good equipment and by such bigger force destroy smaller enemy army to save yet on weapons and people. Then go next to other smaller enemy army group to destroy it and then next. In the end with most rest of army attack main group enemy army if is there. So is usually better than to big army fight each other often without result. By destrucion yet these smaller army groups it may done moral effect on enemy.

One of the best way in case of conflict is to explode bomb in enemy country, even very far with help of sound. When bomb explosion material is lay down in chosen place could be exploded by particular sound, where receiver connected with bomb will act after receive particular sound and then will pull trigger to explode bomb. So in few targets. bombs; radioactive, gas biological etc. be place in few different location in enemy country, then for instance when place close to bomb telephone will make phone receiver go up similar like in recorded message after phone call, directing it to bomb receiver and from own country maybe send particular signal, sound which will be receive from other side and bomb will explode. This way is possible to explode bombs in enemy countries even far away after make it sure they have been hide safely there. Phones better to be destroy with blast for safety reason. So main thing is by satellites or by phone from own country office by particular sound blow them where are millitary instalations or similar targets.

If someone want to receive information from other country, from government or company office some technological or other information so one of ways is to place listening device in building or room in chosen company or millitary units after somebody will place such device. Such listening device could transmit conversation few kilometers away as is already done, to receiver which will be place next to telephone apparatus and then connect dialing by person or automatically to office in own country. This way somebody can hear conversation, millitary meeting, computer sound working from any country in the world about the same time as it happen. So from own millitary office is possible to get information from many countries about the same time on very low cost and profit by this as well.

Some of these listening devices can be self destrucive after they stop their mission.

Also similar way can be place very small cameras sending transission few kilometers away where can be make very fast photos and then fax them to own country or transmit it by own satellite and then to country. It could be done by person or automatically. If not possible to get to particular room next room can be use with ex ray glass which could be place on camera so can see next room by wall what have been done before and this including to receive information from computer. Just have to be make sure to hide properly such devices or make them if need self destructive. Could be cover with some lead to avoid detection or use stronger magnetic device which will not reflect waves. Some trade missions abroad would be use as transmitting points for these devices. Also is not always difficult to receive informaton from computer in other country though such task could be yet more developed to more easily receive desired information.

There is also few methods to make different radioactive bombs. Beside use few known materials like uranium, plutonium atomic particles alone can be excited including done by lasers. Some particles in atom can be excited more than other. By excitement mostly proton it can be extract more energy from proton and produce for example more like proton radioactive bomb, or these recive more energy from from electron and eventually produce like electron radioactive bomb which could be usefull on all kind electrical equipment to use on. Where here fast chain reaction will deliver most energy from excited particles.

These radioactive bombs may have material made or deliver from process of fission or reverse fussion. So when during nuclear process gradually most dangerous radioactive substances still can be usefull, when get to the core instead, they can be tranfer to make ready bomb, mixed thouroghly with volatile feromagnetic gas. Inside wall this bomb could be ferromagnetic or have ferromagnetic rotating shift so plenty high concentrated gas would hold inside. During explosion these electromagnetic walls would be destroy so radioactive gas will kilometers over spread around. To get it more outside after explosion should be create very high temperatur . Yet can be mixed with very inflamable gas to produce radioactive fire for farther spreading. Plutonium is hightly radioactive material in thies case and can do more damage with addition of temperature and smoke. Some such bombs could be exploded few kilometers over enemy army or target with range these higtly radioactive gases over some tens of kilometers in diameter could be create very spectacular wall of fire over part of country. Though it should be calculated how big such explosion could be to avoid such destrucion on very large scale if not controlled.

Precision to hit exact target dont have be so important becouse if target for example will be 100 meteres in diameter when power of destrucion could be over 1 kilometer or more so will be hit for sure. Effect of strenght radioactive bomb with available even simple radioactive material is great. Average radioactive material can be heat up turn next to gas and such gas in very concentrated form put in to small bomb which after standard explosion may release gas at least few hundred meters in diameter. Such radioactive gas can be as mention produce from nuclear power stations., in laboratory and can be mixed with some magnetic or ferromagnetic volatile gas., then introduce to bomb which have wall made with not permanent magnet to attrack more such gas, but after some days some magnets loose it magnetic attraction and have hightly concentrate gas inside.Or could be done with combination very high pressure so very large amount such radiactive gas can be place inside, or use vacuum to place yet more.

Also this radioactive gas can be turn to liquid or solid state so one or few kilograms of this material maybe place to chambers or already made bombs and then heat it up so turn it back to gas.

One or few kilograms not to mention more this radioactive material if heat can turn to very hightly denset radioactive gas with eventual of spread it tens of kilometers or more if desired in dependence of goals, with destrucion similar to neutron bomb. Maybe use any other known method.

Such bombs dont need to be exploded always like in case atack of enemy army or other bigger targets but they can be smootly release after placing them there first. In some case such radioactive gas could be made with substances which have shorter life span so own people can enter enough time later after releasing them. It can be made many such bombs with different design for attacking enemy army or millitary target. Sometime volatile, yet inflamable gas could be put in radioactive chamber to get enough radioactivity and next place it in bombs,. All this some kind is as dangerous as atomic or neutron bomb.

As to virus strategy that no matter if right or wrong they are no better or worst than many other weapons when come to confflict.And is already in some defence millitary program with available strategy and technology. Are in use AIDS but other as well yet can be create only viruses or defective genes which when introduce to organism can cause infertility but dont kill. Can be use some dangerous cancer causing viruses which are able to repair or reproduce themself even in organism.They can be also mix with not harmfull gas and release over enemy even with hire private planes there where can be spread over millitary instalation. Again they are better than some other weapons. By add them to bewerage like soft drink which are often use is other method for spread them with sufficient release by agents which may do between enemy soldiers if carefully done. Have been trials other weapons dealing with fire. There is already innovation which if get out of control could put whole country on fire but on smaller scale is quite sufficient release very denset, volatile, hightly inflamable gas. Such gas is not good to use on no strategical value or wild life just best of course beside enemy army on industrial denset complex connected with it, or eventually crops. Of course such inflamable gas is best to use when weather make air very dry on even on part of country. So by use few containers or bomb in few places can help to spread such inflamable gas over large part area then with combination dry hot air would be done quite of damage on applied part enemy country, especially connected with millitary. This can be done yet with help this or other way of local people as to implicate them in case of discovery by someone when real one couse was in inflamable gas.

If well plan done it may do so much damage from financial point which is also important for country. It could be press as much as possible such large amount inflamable gas to containers or bomb then spread on particular target so would will go in all direction if need. This can be use against enemy army yet this one with heavy artillery where can be spread on fire much of it and more damage than missiles can do by creation on tens of kilometers walls of fire or even against planes if will go very high as well.

It was in trial other kind of strategy dealing with artificial earthquake.

Even if is few ways of doing so this one was done with help of laser beams and some radio waves on different frequency. First was apllied device which predict little early commining earthquake connected automatically with other device which send on the same frequency as was waves coming from earthquake, then when both frequences was on the same device which send laser waves and increase yet frequency making naturally coming vibrations much stronger and increase by this by few degree of Richter scale power of natural waves making erthquake much stronger than original. This all take few seconds and such waves can be directed to more designed place.

Was yet use magnetc field more weak from first side

which detect early erthquake and send waves, beams to designated place where was much stronger magnetic field to increase more vibrations on greater scale to make them similar to original erthquake vibrations. Such lasers, waves, vibrations could be use to these geological points where earth movement are happening by direction to few weaker points to make there move faster or on greater distance. Though mention more in general and need some more sophisticated equipment for such region of country but damage effect could be powerfull almost like atom bomb when is safer to do it and cheaper yet without such bad result as atom bomb.

Especially in areas where it happen naturally so nobody maybe blame when done in industrial part of country that is connect and increase original to other area.

Application other weapon maybe done different way too. Becouse on earth surface is stronger gravitation so is need stronger power to use to lift missile.

Is posiible done in few ways from simple to more advanced from lifting missile or racket with angle 45 degree instant vertical so will be less gravitational power, to three or more stages lifting including to take racket with help few missiles on sides to air, or by planes or even big weather balloons up to 20 kilometers.

Then with one racket still on other fire next bigger one few hundred kilometers.

Then last one which sitting on bigger one could be fire with missile to target as last one. Or maybe use in first stages racket which is use to lift satellites to orbit then on high altitude it could be from it fire missile to target on earth.

Beside as to fuel traditional one like hydrogen maybe push inside under tremendous pressure so lot of it will accomodate inside with material capable stand pressure with addition of vacuum inside too so more will accomodate yet.

When burn even small amount of fuel which will be under pressure and high burning temperature with methods of turning hydrogen to helium will create vey high temperature as well as plenty of energy. Yet again if more vacuum inside then more fuel will get in there.

When large amount planes attack so is better sometime instead send misiiles to destroy each one send missiles with few bombs to explode so by this to destroy large amount planes not just one.

Could be use yet other known methods with radioactive bombs. These bombs not always need be exploded when attack enemy army or millitary instalation - they can be smoothly release after placing them first. In some case these bombs maybe made with lover radioactivity so own people can enter there after some time decontaminate area first also with neutralising beams and waves. It can be make many such bombs with different design for attack enemy. All this may be about as dangerous as nuclear weapons. Above mention weapons maybe made by small number of people with rather limited ability of technology and material which only show how average country can do much itself with all available means on their disposal and if developed consistently made itself more stronger against stronger enemy or even super power by their steady development by add to this own strategy. Though if come to apply these weapons, methods against enemy it better doing and prepare longer and more carefully so this way better result could be achieve by even being notice during such process yet with mention before virus strategy which often is understimate but could be so dangerous though maybe make less dangerous virus just to weaken enemy.

Some these weapons which may weight just few kilograms maybe place in small rakects, missiles close to enemy country or inside just before expected conflict so can be send in few minutes to target. Though these bombs, viruses suppose not be developed but is not known which countries and how many are already develop them. All these weaopns if well controlled and use under limited attack are more safe for world than nuclear weapons but as much powerfull.

There are nuclear missiles which under water in ocean

can by blast create large waves which could destroy costal land, but these same could be too use from other side directed opposite direction to neutralise these waves.

Instead of bombs and missiles could be use if is no other way for defence to send against enemy army antrax viruses one of most dangerous and after conflic finish radiate whole place so can be repopulate again.Also in such desperate case may come to use ingridient from group dioxins called 2,3,7,8 tetra chlorodibenzo p -which is many thousand more stronger than average poison, so very small amount of it if put to bomb one of describe ways and later exploded can spread around very far make it not less dangerous than atomic bomb but safer later and they dont do material damage. Such kind of weapon may be use when no other way of defend itself and want to save own life. Later of course have to be clean whole place where was contamination for safe pass.

By application mentioned sound principle can be done simulation of laser beams which may look like they hit target anywhere on earth from own country.

Though such real laser gans have been invented with short range of destruction but simulation may give impression of these real one. These known lasers can hit and destroy target in distance few kilometers. So one or few laser guns would be located in hidden place in other enemy country or few countries or close to their borders. It can be directed to choosen target in laser range distance then when fired of course can destroy this choosen target. But here nobody have to be present, just especially prepared device may push automatically bottom or switch it to fire laser gun. So by the same principle as formerly done by telephone with sound to blow up bombs, here by dialing number telephone which will be close to such laser, even mobile telephone, speaker will be directed to receiver in device so with particular sound known only to these who doing such task it will switch or fire laser gun already dirrected or automatically directed on target.

So for instance somebody in own country, city can dial

known number in other enemy country where have been already place these laser guns and then by play of particular sound to receiver switch device to fire laser gun to choosen target.

To make it more properly working can be make call few minutes early and next in desired time just can be push bottom to play such sound which will move mechanism to fire gun. This way target anywhere in the world would be destroy in one or few seconds. Similar way can be done to ships on sea from any other not far away ship, even commercial one or submarine. Or by placement few laser guns in different countries and make call from few telephones can be then destroy few targets simultaneusly anywhere. To make more real could be also invated people from other counttries in dependence from circumntances and when dial in secret few minutes before actual presentation it can be show that just by push of bottom can be destroy target in any other country or sea about the same time.. So when push bottom in present of people the same time will be fire laser gun in other country or own in very far distance or in ship. But have to be done properly so nobody would discover that it was done by place there laser guns which later can be remove or destroy by self destructive device. Al this can give impression that country have real laser guns which can destroy target just by send beams straight from laser guns to air. Also in case of conflict this could be apply to one or few enemy countries when expecting attack from them.

So it would be like kind of warning.

When sending missiles or rackets over enemy country to make them more invisible to radar may cover them on surface by thin but strong magnet or electromagnetic field which would attract waves send by radar. They are many kind of waves though few basic ones in use and therefore can be make one more universal. Also metals and waves can be magnetise so can be make electromagnetic set on surface to send strong magnetic or electromagnetic waves when over enemy country or close or have device inside to send on surface electromagnetic waves all around. Some bigger especially intercontinental rackets or planes can contain small but sufficient device made in laboratory working on implosion process similar to this in black whole where all waves around are pull inside, but here connected with electromagnetic surface to absorb all coming waves. In macroworld some part of this process can be seen in air conditioner but whole process here better take place in microworld in subatomic particles. Similar process in implosion could help in production nuclear bombs and then fast reverse process to explosion including by change of magnetic field

In more simple way rackets can be make or have very small radar so when detect coming missile aim to destroy rackets then by small electronic device which send signal to own smaller missile set on rackets to shut coming missile. Like for example when enemy send missile to destroy own and they been shut down, here with small attach radar it would detect coming missile and then fire own small missile but with high explosive power which would shut down coming enemy missile.

Also racket could be inside shell so after detection coming missile shell will open and racket will fly farther when original shell would be destroy.

Could be shell use too when racket come from outer space to save on higher temperature racket and maybe use too electromagnetic field in shell stronger from inside to push against other wall of shell as preventive. Maybe few these shell not only one use. These racket can have few very small missiles with them detect by radar of coming missiles-these few missiles can send to target living original one which could be destroy. Yet when attack can be send from transmiter radio waves to enemy millitary,radio, TV stations to make there disturbances or give own wrong messages.So not only have to develop long range missiles but these one with strategy can do the same.

However there are better and even more powerfull weapons with strategy of use of it which could make nuclear weapons less important. They are best for country whch have developed space program.. When send rackets to the space on the way happen obstacles with asteroids there. However here these rackets can be attach to small or bigger asteroid or few of them and next direct it to the target on earth. It could destroy any target including nuclear weapons hide under ground as impact of asteroid will go very deep inside ground.

With very small asteroids could be use at least few of them at ones to attack enemy country with not that big area of destruction. If nuclear weapons will explossion be done from underground it can be about the same time hit with asteroid to make less radioactivity around or done with few small asteroinds.

Is also possible to send asteroid like rock just from earth to high altitude and then down hit target but speed must be also very fast to make impact similar to this one from space and can be increase explosion by place there inside explosive material. With proper calculation they are more powerfull but more safe than nuclear weapons as dont spread any radiation around.

From better side of use such asteroids when attach to them rackets or missiles they can be directed to other bigger asteroid which come closer to earth so could destroy it or push back. Few these smaller asteroids can be use if necessary at ones against bigger size asteroind. Racket itself which move asteroid dont have hit ground on earth just asteroid itself when racket can be destroy in atmosphere. All this with proper calculation can do big damage to enemy country but destruction could be much less done like for instance nuclear weapons.

However if dealing with other weapons maybe mention that in present is sometime better to apply instead of intercontinental ballistic missiles of small supersonic jets. Some small one have speed over 2000 kiliometers per hours and also some small supersonic jets which was in sale and can be made by somebody too, have speed over 3000 kilometers per hour and with improvement may increase speed to 4000 kilometers per hour. These super-sonic jets are better sometime to use that ballistic long range missiles and with improvement in engine and resistance to higher temperature and air pressure will be like very long range ballistic missiles. Some fastet jets like X-15 have already much faster speed not to mention X-30 jets which if developed could fly around earth almost in one hour. So in such case would be better sometime to use them instead intercontinental missiels with improvement and development.

They could fly with small missiles attach to them and shout when in need.

Missiles or planes beside can fly on chemical and electrical power with two different engines so may have farther range and speed and weight less.

Beside if want to develop them and some other weapons dont need always to do on own as already are many advertisement about not only these and other weapons but also about important for science and economy development.

Very often they are ignored and becouse bussiness is often too more important than human live they will be never developed. So is often better to check in around about these new technologies from companies or privite individuals to be in development ahead against other countries.Is better also to check and try hundred different ideas and find out that one or two are working than just abandom them becouse may seem to good or look to far away for this time.

More energy can be achieve by application antiparticles. Average atom could be more or less turn to antiparticles or anti atom by radiate them with properties which are opposite to them. For example radiation which can change magnetic moment to opposite, or change electrical charge and other properties. This may be done by application from opposite charged atoms, or opposite gradually change proton to negative also. Such gradual radiaton particles of atoms would turn them to antiparticles. Some help could be also by manipulation with more or less temperature, or application magnetic field. So by this atoms will move more slow against former speed. Also can be done general radiation all field where atoms move so all that space would more resemble antimatter space close related in physical properties to change already atom properties.

Around could be create small field, even rotating which less little resemble antiparticles but but enough not to annihilate them. Then next small field (also can be rotating) would be created even less remain former field and last layer of the field more closely remaind normal particles. This would help create different properties of atoms turning around in their own field without annihilation becouse external contact. Creation of atoms with opposite properties is slow process but is possible to do it by use mostly radiation delivered from these atoms particles which are opposite atoms to other atoms particles in physical properties or opposite atoms to each other in that respect.

During bomb explosion these small field will be destroy and when these anti - particles will come to contact with will normal here field, air- great explosion will happen. This may have too positive effect. For example if radioactive particles will come to contact with antiradioactive ones they will be annihilated each other so negative case of radiation may be prevented.

To make sure when transfering radioactivity from nuclear power station or laboratory this radioactivity first also could be magnetised, which can be done by magnetisation radiactive material itself first, so that during transformation more or all radiation will be place inside bomb which walls can be electromagnetic etc. or as mention with help of rotating electromagnetic shaft inside bomb.

When mixed with magnetised volatile gas then radioactivity will spread far, but if mixed with more heavy elements gas will spread less. After this explosion later can be done few other not dangerous which would contain any safe magnetised gas with no heavy elements which could go deeper to the soil attract less magnetised radioactive gas. So this may clear area faster from effect radioactive gas, even from failure in nuclear power station. Also as menton before can be also spread on ground very thin magnetised sheets which would cover large area on ground around to attract radioactive gas. This could be apply in nuclear power station where this radioactivity maybe magnetised first and in case of break there this magnetic sheets would cover leaking area and other magnets around will also attract escaping magnetise radioactivity. Is possible also to apply this to biological organisms.

Almost none of this known technologies can match speed and maybe some way destrucion done by laser beams and radioactive laser beams, yet may be with combination radioactive radio and television stations. Though not any material damage will be done may could destroy any country or countries to large degree in very short period of time. That is why better to use if in need to some limit as only one or few person with radioactive laser may do damage to large target from very far distance, even to plane by radiate pilot or crew. This would destroy only biological organisms but to destroy missiles, planes or ships have to be use other principles.

With present technology laser which would destroy target will shoot on very small small distance away becouse lack of power. But can be use standart laser which can send beams millions kilometers in space. Such laser yet could be magnetise by few known ways already. Other laser which can shoot only on short distance away also could be slightly magnetised but much less. If such slightly now magnetise will be place very close or almost touch with mmore magnetised standart beam will run, travel with it also to space even millions kilometeres away.

But becouse now short laser beam could be much spread or dont have so much power so will need few laser connected together to main laser which send beam to space, so all beams will be in touch together. With few such beams together shooting in one target especially at least few seconds it will bore whole inside or destroy target. So standart laser which could send beams to space have to be some more magnetise and to it connected few lasers less magnetise of short range to hit distance., so their beams when stick to more magnetised standart long range laser will go to space with it. Few these big lasers could do damage to planes, missiles, ships or by satellite hit any target on earth. To make this more effective it could spin around fast when shoot so will bore whole deeper in target.

There are also other ways to do it. Laser could be made with very strong acid resistant material. Then some part of it made with very strong acid material so when such beam is send to space will be also very acid like which could burn even strong material if hit target. Becouse to hit target on earth will take just part of second this strong acid or very hot as mention before would not destroy laser beams as is to short time for it. Such acid when made can be mixed with vey light ferromagnetic substances or slightly magnetise with resistant to acid material or only radiated so such acid would be place close to beam and then travel with beam to desire target.When use such acid gas or substance like perchloric acid or even stronger in dependence from target and acid used it will burn target after a while and radioactive could be add if necessary.

When using laser over enemy some lenses can be use to spread beam around over wide surface. Though impact will be more weak, but if use radioactive beams, poisonous or their combination if done by longer period effect will show up.

Is good to use lens when beam pass by mirror so when concentrate to very thin point it will make hotter beam from awarage beam inside laser. Some part could be worm too to make little more worm beam.Yet few laser can be concentrate to the same target at ones. Radioactive with poisonous waves is possible to make in accelerator and when they reach highest speed direct them inside laser to send them with laser beam to desire point or send also just protons extracted from hydrogen. Some antimatter is possible to create in laboratory to start with antti protons as some is already done in laboratories or antielectrons; positrons.

Such long process maybe help by radiations on electrons to make them positrons or general radiation which change character of atoms to opposite with radiation with opposite characteristics, or partly with help of cosmic rays or some their other cosmic coming particles. Even if such radiation will be not completly change all characteristic to opposite still will have some destructive power when meet opposite matter. If these beams introduce to laser their inside parts have to be more compatible with such close antimatter and send to the target with other lasers send the same time to the same target. When collided would make explosion which power depend from greater differences between matter and antimatter.

When search with microscope for smallest particles may help more if gamma or roentgen waves will be use to give better more clear picture. They rather dont bend but if slightly magnetise and then in corner of lenses of microscope will be use small magnetic mirror to direct them inside they may bend to these as in strong magnetic field they bend when first yet been strong magnetise in such field.

Some other magnetic combination maybe use as well also with magnetise lenses and similar way use in case strong bending waves which normal way would not get to lenses not matter their angle but with help some magnetic combination they would not bend that much to show much better picture and with help of mirrors can be transfer on big screen for more details to show. Many particles or subatmic particles cant be seen becouse they are smaller than waves lenght. But these particles no matter how small emit also other radiation on which with application some colours can be seen as they emit other waves like includng sound so with again application some very sensitive micro sonar apparatus these sound waves will help to localise hidden sub atomic particles.

Yet if will be use special diafragna etc. to increase or transfer received sound.

After their localisation is possible to use radiation or laser by longer period on these particles to change their characteristics. Proton for example to antiproton by negative radiation. As hydrogen have protons is more easy to target and change characteristics this atomic particles. They dont have even to be radioactive just standart hydrogen by turn to radioactive by radioactive radiation with opposite charge to turn them to antiprotons against former ones. Later with standart proton as opposite it can be use too in missiles as fuel becouse even small amount would create lot of energy. If use hydrogen itself with oxygen by some ingnition will create explosion. In large amount small chambers introduce hydrogen and oxygen could create small explosion and the same time these large amount small explosions would have big impact power. With help of accustic which will increase sound many time over original power of explosion and sound explosion will give more trust to engine so even small amount use to protons and antiprotons or antimatter with help accustic material is able to create additional power save on amount of fuel. Similar principle is possible to apply in laser sending sound waves to very narrow point or to create sound bomb with explosion increasing sound many time more directed some way. Beside combination hydrogen and oxygen create bomb explosion send by missiles with ingnision to make large fire around. So with many missiles it would take large surface under fire which will do more often damage than standart missiles against army, ships when yet poisonous or radioactive gas is add.

There is few other radioactive devices which if developed will be much more effective than intercontinental missiles., or planes. First have to be build radio stations which can send radio signals to every radio receiver use by enemy, including mobile phones and similar devices. This radio station or few of them may have electricity source which will be higthly radioactive. Also some parts radio stations with transmitter antena also will be very radioactive, so in case of conflict if know enemy frequency these radioactive radio stations can start transmitt signals to every enemy radio receivers and mobile phones, in planes, ships and in land. In matter of seconds it will cause radiation to the body with no of course matterial damage. At present time armies are well equiped with means radio communication so if this will done on whole army of enemy most people will be affected as from mobiles and other devices radiation will spread too around. However becouse at present this technology information should be rather limited becouse if will be build radio and television stations which will this way have radioactive electricity source and some radioactive parts, including radioactive transmitter antena, so then would be possible send to whole country or even all countries very strong radioactive signals to all radio and televsion receivers in pick hours, also to computers, internet and by sattellites yet weather stations all over country too also run by computers power stations, so most people in enemy country would be affected in seconds and if this will continue by longer period radioactivity will be spread around. Becouse of portion radioactivity over 3Gu can be deadly so have to be calculated how strong radioactive transmition have to be. Ultraviolet light is very dangerous to human organism. Likewise some parts of transmitting stations maybe made of plutonium which is also one of most poisonous substances. Some parts of radio and television stations could be build under ground for safety, just mostly antena can be outside high. This is much faster reaching than any missiles and dont cause matterial damage again.

They have been going research on so called dead rays which can destroy on distance and also Star Wars program which could destroy enemy missiles very far in space.

There are few ways to do that or similar. One of them is use average laser which can send beams as far as beyond moon. These laser could be run on batteries or powered by generators. But like in former case batteries or generators maybe made very radioactive, the same like some parts of laser including mirror. So now when laser will send beam away such beam will be very radioactive with degree and kind radioactivity depend from radioactive matterial use and it amount ; uranium, plutotonium etc. By use optics beams can be spread farther around so when enemy army attack by use few radioactive lasers whole army maybe destroy without any matterial damage and from far away. These radioactive lasers could be use against ships or planes or from ships and planes.Also by satellites or using transmitters in sattelites, weather balloons, planes. Radioactive beams can be send over any target When spread dont have much strenght but by longer period and few of them radioactivity will show up. If bigger laser bigger effect will be.

Is possible also to move very small objects with speed of light that is about 300 hundred thousand kilometers per second. To this could be apply few ways.

One of them rather simple is by application of laser. They alone can send electromagnetic waves with this speed. However in laser generator flat mirror from which electromagnetic mirror polished to the same degree as normal flat mirror so far use ,or little better, just little, slightly magnetic. Also some other parts laser generator also could be made with some magnetic matterial or slightly magnetised. Beside source electricity could be too magnetise or so maybe solid or liquid gasenous substances which are use as laser materials. So one or few mention parts can be magnetic or magnetised in dependence from how strong magnetically electromagnetic emission (light) is desire to be. Now when sending beams to space such beams will be more or less magnetic. Just important is not to much magnetise part to hold beam. Before mirror by which beam wiil pass can be lay very thin like razor, light piece of ferromagnetic metal. Becouse beam now is magnetise this small, light piece of metal which also could be magnetic or laser ferromagnetic- will be attract to magnetise beam escaping from laser so by this will be send to space travelling on front of beam. It will also help becouse magnetic attraction. Focus beam in one point to this small ferromagnetic piece of metal instead of spreading it.

Beside small ferromagnetic piece of metal could hang in air in some distance from laser hold only by other crossed lasers beams or any other known means so traveling magnetise beam will push this ferromagnetic metal ahead.

The stronger magnetise laser beam will be then more heavy ferromagnetic piece of metal will be able carry to space, and also if bigger laser generator then will be able to move more heavy object too. Just if is need speed of light then laser have to be big enough to carry such task, otherwise speed will be slower.

Beside that maybe use more than one laser to move object to space. When one laser will be focus on one point where it will be send and simultaneously shoot on small metal object to carry it to space. More these lasers and bigger with stronger magnetic beam then heavier object they can move ahead or any way.

Also object dont have always to be carry in front of beam only when shut beam from different angles straight all in the same point they can carry object wiith them after they are almost cross to each other and then send straight ahead. So object will be more protected in space inside beams. Such lasers could use different frequency when carry object from invisible to eyes to normal visible beams. They can also carry non metallic small object which will be wrapped in thin paper like magnet then put inside equally thin magnetic box which have walls made from opposite magnetic poles to magnet- use to wrapp non metallic object.

If magnetic walls of box will be many time stronger than magnet use to wrapped object then becouse of repulse magnetic power object will hang in air inside thin magnetic box never touching walls and by this will be possible to move more heavy objects with laser beams. But even very light object with such great speed could do damage to target where was directed.

So in short it would be describe that hightly radioactive gas or other radioactive substances could be strongly concentrate inside bomb mosttly by means of magnetisation and pressure or very strong magnetic field to direct explosion inside first few layers becouse opposite poles of magnetisation.

To magnetised radioactive matterial so later in nuclear power stations or on open space have to be use strong magnetic matterial by means to attract such radioactive magnetise gases radiadion etc, to lessen spreading radiation.

Or to magnetise laser beam or matterial source so any ferromagnetic or magnetic object, particle attracted to beam so move with beam, beams and also calculate object weight, size of beam, beams to maintain light speed.

If want move object faster than light speed have turn around laser beam so small object attracted to beam by means of magnetisation will move around 180 or 360 degree accordingly. Even in space -so when for instance laser beam with small object will turn around 360 degree in one second so beam which was already 300 hundred thousand kilometrs in space this 360 degree around will cover over one million kilometers, or if turn around in few seconds then accordingly will move yet much more on distance.

To connect few short range destructive laser beams ferromagnetic or slightly magnetised large long range standart laser beam stronger magnetise means spread for long distance short range but destructive beams.

To made very strong resistant to acid laser so can shoot strong magnetised acid or magnetised acid particles.

Inside very light box air pressure will keep heavier object in air so will not much weight contain and shoot magnetised laser beams to carry object in light box by means of time saving between target hitt and object inside box falling down.

Radioactive waves which are made from radioactive matterial or by means radioactive parts can be send by laser or transmitter any kind.

Very light magnetise object in front of beam would be attracted to beam when move trough atmosphere so dispersion of beam is prevented.

By then is possibility that from space laser beam will be

directed to very small magnetise object on earth and then move such object very far even to other part of earth in few seconds. Like for example very thin metal object lift up then move from one continent to other. It is rather simple thing to be done without much technology beside this known now.

If small explosion device will be place inside little magnetise gas so with end of laser beam could be move to very far distance or other part of earth – with few these explosion devices shot one after other or with few beams they would make effect on enemy territory and army. Similar way maybe set inside such gas hot plasma which would make lot of damage as well even against planes or ships yet other objects. It was make trials on shorter distance but for any distance principle is the same.

In middle of beam by small mini pipe in laser is possible to place hot piece of gas with vey hight temperature or send the same time hot laser beam with a short range so when again travel inside beam to target it will burn surface of target.

Yet few these lasers fire simultaneusly in one target will make bigger effect.

Electrical charge in small portion would be too introduce inside of laser beam which is sllightly magnetise then usual or small portion ferromagnetic gas first or low pressure gas which is good electrical conductor, then portion electric charge and fire simultaneusly. So this way will run to the target with some higher voltage.

With yet many lasers very fast firing whole army maybe effected by some of electrocution from very far distance. Also use against electrical equipment and even radio, television and other receivers. This may cause short circuit. Radioactive gas also maybe introduce inside beam this way. Though some parts have to be isolated agaunst case of electrical charge, Beside this way can be send luminous beams and substances which effect senses of taste and hearing.

To make beam not to spread when passing by atmosphere before mirror should be placed almost transparent and almost weightless but strong enough magnetic object so less magnetise laser beam will be attracted to this magnetic object when run to space. This could be use in other lasers where beam spread as pass especially by atmosphere, so with very light magnetically strong material beam will be attract to it when only slightly magnetise itself. And again if two or more lasers will be focus in the end of beam together yet with slight magnetisation it will stick in the end of beam together also.

With all these lasers using describe principles and radioactive radio and television stations is posible in short time destroy all enemy countries or choosen one yet most of course millitary targets, if this is developed.

Two of these devices that is moving small objects with light speed and long range destructive laser beams or radioactive are basically the same principle devices only some ways of using them are different when nuclear weapons kind here describe how to make them so when radioactivity is create or radioactive gas then becouse exceptionally smal weight they can be magnetise very easy to carry by magnetised laser beam to target anywhere on earth and they will have similar effect as neutron bomb becouse only biological life will be destroy. However it will take only part of second to send them to destination to any point on earth with help of satellite or many stationary weather balloons if locate properly target.

That is why if these weapons will be developed they will make other known much less important mostly becouse of speed but also range of application of it and effect though they are less dangerous to use that known nuclear weapons becouse if controlled they may do less damage and no so long lasting.

Of 142 atoms of uranium only one belong to lighter isotope which can be use for bomb explosion though important is to have small surface of uranium against its size so less neutrons will escape. Critical moment come when will be more neutrons than these which have been lost outside parts so would not hold longer their accomodation. Though it will help sometime to rotate to hold neutrons or slightly magnetise uranium with more inside part and create opposite magnetic field outside. By combination can be create even proton bomb. By losing electrons and pushing atoms together is chance to have very denset matter which even with very small size would have exceptionally big mass or weight.

By magnetisation with rotating movement it will help, or with condition similar to these in stars create in vacuum with very low temperature denset hydrogen and spin around. Star spin around and rotate in space often tens km. per second, plus yet long time of process. As result of high temperature and very high pressure inside atoms are losing electrons and becouse of rotating movement atoms are get closer to each other, plus hydrogen still change to helium and if is not means to escape neutrons will get to critical mass and explode. Also in space stars are penetrated by cosmic rays much more than in laboratory. Here will come to help collision deutrons and trithium which can be obtain from sea water also and trithium is available too.

At present to send missiles of medium range few hundred kilometers is not much achievement. Similar result was already done during second war with V2 which weight average about 12 tons and travel around 359 kilometrs with speed usual 6000 km. with fuel which was alkohol. Today becouse of improvement will fly much farther yet when most oxygen could be take from air. Smaller piece of uranium could be source of power with farther development yet when mention before ways of use moderators will be apply which solve weight problem.Other problem is that material which is throw out will be very light so would be not enough to move racket. But if it would be throw out by very small micro holes yet with very norrow ends so would make denset matter pressing by such small micro holes.

Sensitive blades, some diafragma would move just with pressure concentrated sound. Here pressure on blades can make them move very fast and blades could be part of small turbine which maybe connected to generator which will power electric motor. Could be made combination of traditional fuel to start and fly in of use uranium power.

Start maybe done in high mountain where is not that much already denset air or higher from supportive by altitude strong enough with many of them to lift racket. Beside electric motor could be powered by high voltage batteries, lot of them in vacuum chamber for less weight connected to motor. Or yet use traditional fuel to start and electric power too later. Uranium or hydrogen can produce much of heat as well to use moving small turbines to produce electricity. Standart fuel have enough power after burning to spin small turbines for make electromagnetic induction as next power. Traditional fuel like hydrogen etc. maybe push under tremendous pressure so lots of it will be place inside with material capable stand high pressure and from other side of matterial in very thin space also could be push more fuel under pressure for balance to first pressure inside. When is burn even small amount of fuel which get out will be under pressure and high burning temperature with method of turning hydrogen to helium will create very high temperature as well and plenty of energy. Is possible to pressure such fuel to vacuum if more vacuum space more fuel will get in.

Matter of mention is that lasers can be also use to send alfa, gamma rays and other elements from atomic particles which maybe radioactive or just send these particles which can destroy atoms in biological body so enemy will be later more biologially weak and more easy to get infection becouse of weakness and in such case would be sometime hard to trace original source.

Instead of air planes yet for countries which dont have them much, yet these millitary air planes, would be alternative to use in defence stationary weather balloons. On the long border line just before maybe place many or at least tens of these balloons with set there guns, missiles to use with remote control device. When on radar will be seen coming enemy planes or army on ground then by remote control these weapons maybe use to attack enemy.

On higher altitute can be seen far away distant picture of ground. Also these balloons with cameras could be use to observe own border and whoever will pass by it 24 hours if necessary.

For energy purpose is sometime better to apply different chemicals or minerals, for example lithium which when bombared by protons could give over 56 millions calories when coal in this about 75000 calories. Nucleus of uranium when split then one gram of it would give as much as three tons of coal.

Though in nuclear reactors can be use also uranium salt or thorium which have similar radioactivity. In some case to fasten or increase nuclear reaction in uranium could be use accelelators where some particles when reach highest speed could be direct to nuclear matterial so would have similar result like bombardment by cosmic rays. In dependence from goal maybe use electrons, protons, neutrons, positrons, mezons – which are too in cosmic rays or made in atmosphere when pass.

Cosmic rays are more penetrating and have many time more volts than standart radioactive elements. Also radioactive particles maybe directed from accelerators directly to the laser and send together with laser beams which will be radioactive then as well. Atomic pile is even better yet to prevent escape neutrons from uranium as use moderators. Few matterials help in this, including even pure thick wall of salt which too hold much cosmic rays and they dont send gamma rays.

Some hard minerals could be made to thick gas and use as moderators or against escaping neutrons, as thick gas they dont weight that much.

Was use also way by rotating uranium so by this hold inside neutrons. Beside was too use way rotating moderators so many escaping neutrons jump back.

Though speed was regulated as for some reactions are need slower moving neutrons. Yet both materials maybe rotate. Other way was slightly magnetise whole uranium and moderators was made of more magnetic gas or rings and seperated outside so these magnetic moderators have opposite poles. In strong magnetic field even cosmic rays are bend to some degree.

As is known in genaral to make atomic bomb is need to seperate isothop U235 from U238. Uranium is different from other elements that when bombarded by neutrons its atom split on two parts and its atom split on two elements for example krypton and barium. When one neutron divade one atom then chain reaction start, more neutrons split more atoms and process will continue unless beside find way to hold neutrons. If bigger piece uranium is more easy to have such reaction.

To prevent escaping neutrons could be use some mention before methods of use moderators and ways, it would hold much more neutrons inside.

By this will be possible try use smaller piece of uranium in laboratory scale with similar way of doing process.

Uranium not need to be bombarded from outside as will be enough own neutrons and cosmic rays. To trigger bomb will help fast connection of smaller pieces in one whole together. In hydrogen bomb process is different and have to be made atoms of bigger mass from atoms of smaller mass. For example from two atoms heavy hydrogen make two atom of helium. Yet in hydrogen bomb may start synthese not from heavy hudrogen but from isotope of hydrogen call tritium. However reaction synthese of helium need so high temperature that to trigger it is necessary explosion atomic, uranium bomb.

In case using uranium it have to be pure enough to get proper result.

And during split of atom have come number of neutrons smaller than three but bigger than original.. This as was mention is known in general.

Yet in these process of making is better to use vacuum space as much as possible so process reaction will go faster then. In case hydrogen bomb also is possible to have stronger magnetisation inside parts and less outside so procees of impulse will go more inside. Yet in atomic bomb reaction and process have to be reverse.

And of course mass of every atom is never the same against each other becouse is change when in motion.

When more matter will be place in smaller area then usually is more mass or weight but if gravitational force will be not only from one side but also other side, sides then mass,weight would may change also. When yet the same matter with the same mass, weight will be measure on earth and then move to hight altitute or to space so mass, weight will change even with the same amount matter in the same area.

Also in macro world mass of moving or throw object, body is directly proportional to force which move such object, body. Though have to be consider other factors as well. Like outside forces, air, gravitation etc. If with the same force object will be move in water, then in air – so in air will go farther with the same force which move it and in outer space even farther. Yet will growing more and more vacuum will gradually move farther and farther with use the same force.

Now does mass moving faster and faster object have mass bigger or mass is the same just have bigger and bigger force which is set in other object which move first one.

If in laboratory will be done trials and in accelelators with smashing atoms then is possibility of creation black holes. In future with improvement these black holes may become bigger and more dangerous. But in such case have to be create even some bigger force – like type of neutron star for instance which could destroy growing black hole.

As to other subject, alfa missiles could be use also to shoot some desire target which on earth condition here they will reach target very fast. They could be made more denset similar to other beams, also in laser so will have bigger impact when reach target. In the end when coming out would be very norrow point so more alfa beam missiles will be accomodate and make more mass by that.

When make application by very strong magnetic field is possible to slow down atom rotation especially if is done more in vacuum. Yet then if will be done with rotation movement opposite to the atoms movement or electrons. As between atoms themself often or nucleus and electrons is still far distance in micromeasure so when some atoms will be destroy it will show up these yet smaller particles from which atoms are build and is chance by this to receive signals from these smaller particles.

It is difficult when bombarding to find nucleus becouse of small size. Some help maybe done by slight magnetisation hydrogen and then send from it by radiation to atoms, then with yet smaller magnetisation alfa or protons particles to bombard hydrogen atoms so still its smaller nucleus may attract these alfa or protons particles. If bigger vacuum then more easy alfa, protons will pass.

From other side by influence strong magnetic field is possible to slow down atoms division, including in uranium bomb.

In other case if will be apply from outside by strong magnetic radiation, waves to the place where nuclear weapons are hidden by longer period it can stop division of atoms and is possible to destroy them in next stage by for example application very strong vibration even on micro scale similar as micromagnetic radiation or waves would done. If electrons or even atoms rotate in particular way then radiation or vibration have to be done similar way and as much as possible similar rotation speed and increasing vibration as much as possible.

Of course to slow down vibrations one way is to increase vibration speed movement around. Also radiation can be lover, done by vibration atoms until they will be destroy. Have to be calculated electrons and atoms spin to the frequency of vibration even to micro scale.

If atoms will be destroy to smaller particles then with help magnetic radiation they can be rearange if done more in vaccum space but still in some magnetic field if necessary.

To create more vacuum space are few ways. One of them is pull, pump out by traditional way and then slightly magnetise remaining space and atoms. Then with yet stronger magnetic field out try to pull out as much as possible remaining particles and elements which are there. It could be repeat this few time to create more vacuum space. In such more empty space hydrogen atoms will be more seperate from each other. But if will be pull out space between them hydrogen atoms should should get closer to each other and eventually press each other more and more. In reverse course they will move from each other so gravitation will be more weak too. All bodies are attract to each other but this also depend from general whole space gravitation which is like kind very thin elastic blanket keep all bodies together with some disturbances, sometime couse by speed of bodies, their mass or other forces. Objects in this space are created becouse still large amount creation of atoms especially hydrogen. If creation of atoms will stop then will be no more new objects in space. And this depend from yet more inside field much below atoms which eventually will become more weak by slow down process from yet deeper field for creation of atoms.

Cosmic radiation is much stronger than this artificial made so fat by technology becouse it come from many sources in cosmic space. So it is not homogenous and comic space all together have very powerfull field. Then to increase artificial radiation also have to be apply many sources of radiation and create as much as possible strong field from which this radiation will be send, not just from one source. Yet these created objects better to be similar to these in cosmic space.

Cosmic radiation also consist positron which if will be use to bombard hydrogen nucleus will throw out electron from orbit. With greater speed when bombarding nucleus with alfa and other particles will create bigger force to struck nucleus.

Such faster speed to achieve could be of course by use vacuum chamber so alfa and other particles will have faster speed becouse less opposition from matter in chamber on the way.

When send light waves by very small hole in board then from other side light can be spread becouse when move make pressure on matter which from other board site have bigger mass to hold more light and also matter particles movement.

Different if from other site of board will be water and yet different if will be there stronger magnetic field. And yet little different if these light waves will spin very fast or will have slight magnetic character and other site will be yet magnetic field with opposite sign to repulse waves. So all this depend from matter condition and forces by which light waves or other are passing.

Dependence between electrocity and light is obvious as also between other waves and forces. When atoms been created start from hydrogen they start too send all known and not waves forces and particles too becouse atoms themself was create from yet smaller particles and these between atoms and yet their movement.

Incidence angle should be the same as reflection angle but if from site of reflection angle will be stronger magnetic field it could bent somehow waves so reflection angle will be not exactly the same as incidence angle. A sometime in water or water drops when move they bend too little light waves not so even.

This coudl be apply in microscope still to increase size searching object. Yet if all will be done in chamber where is as much as possible vacuum not only in microscope then is still chance to receive still smaller searching object. Do not mention that by sorrounding matter there waves will pass faster and deeper.

As was mention too before this magnetic field when use as atoms was destroy to smaller particles so with application magnetic field and micro magnetic force to rearenge these parts to desire shape or something different -similar like in macro world when by magnet is done sometime rearegament small metal pieces to desire sahape.

With modification alfa and other invisible waves maybe use in radar to receive these out of range to find by radar objects. Have to be yet find next still unknown waves which atoms and particles between atoms send. They are without doubt and one of proof of them are some animals which have yet other senses than humans and some of these senses receive these unknown waves, vibrations.

In air planes and other flying objects when they fly yet very fast is difficult to turn them around, on sides or up down and yet farther away. Then with improvement could be apply magnetic devices, seperate so will not disturb other devices and then these magnetic devices should have on each side opposite signs when flying with fast speed can be switch for example from left side to right by electromagnetic force which from one side have opposite sign to other so with enough strong repulsion could turn flying object to right side much faster. Or maybe use both side of object like plane to turn it one way even to some distance, or up or down. Maybe set few sides engines which after such side push and disconect main engine would yet faster move flying object to desire side by electromagnetic push.

Such flying object maybe connected together assemble with few parts and even when flying fast these few parts still maybe connected together each with own engine. When need and still fly these patrs may seperate to two or four parts and each fly with own engine in desire direction. Still yet during flying again maybe connected together. Persons inside dont have to be affected with such fast movement, mostly on sides as maybe apply around them magnetic field, walls with opposite signs which would keep them in the middle yet if they have some mettalic ferromagnetic cloth or could be apply air pressure from each side to keep them in the middle in air so even very fast side movements will not effect them. Something similar would be use in passenger planes so when crash people in the middle of air will have more chance to survive if these devices would be swith on in right time when needed.

Even in outer space astronauts could use magnetic shoes when walk on metallic floor to keep them stronger to this floor. As magnetism still have many use and come from atoms movement during first creation so proton pull electron to itself close so by this probably magnetism was by that first than electricity by micro seconds.

If hologram outside screen will be cover with very thin metallic magnetise powder would may look as more solid object not only hologram. If then will be send by waves or laser beams to high altitute over enemy country may look too as solid objects are coming. Is easy show reflection on wall from small mirror, but can be put few small mirrors one under next and still will show one mirror. But if seperate will show running few reflection on wall. Now if send to space also as hologram on high altitude could show one object fly but if by hand or automatically they will seperate as mirrors on wall before may look that from one object come few seperate objects.

They also the same way could be bring back to look like one object again.

All space is one big field made by atoms and macro bodies. If in big room will be place small balls, metallic or other and some cold orther hot, then start to rotate them with some small motion and other very fast so next eventually whole room will become fiill up with many different forces in dependence from what kind what made of atoms these macro bodies. All this will create gravitational, magnetic, electrical and other field which by that will keep later object more close together. Sometime becouse of mass and speed particular object can break this field and travel by it or in egdes could get out of such space field. Something like in oacean in which later would be this kind of field in comparison in which all kind bodies, objects move, swim. On large cosmic space is similar with greater much force.

Atoms if have more space between them would move more freely but not so in more denset place with bodies like in ocean where water is this kind of field in which bodies move. So in room bodies from atoms to big objets create such flexible field fill up with different forces and which connect all these objects. So by this field and by forces in it can be send signals from particular object to other on any distance just speed of signals in limited by these created forces in field and matter which is still denset for these signals. If atoms are more seperate is more space between them so signals can more easily pass but for farther distance there will be always atoms on the way.

If vacuum will be created could be use magnetic field to attract them but this magnetic field can be make flexible to streach more with atoms in this field. Streach as well until they become transparent in comparison and still streach this magnetic field more. This may include radioactive atoms so they become less radioactive. Is also possible push such magnetic field back still by use opposite signs in ends of such field and add to it other necessary elements during such process. It could be done in this field on micro or macro scale.

So then if signals are send by space from one or ther object it will in time pass to every other object in space even very long time though in the end will be not that much strong as in when start. Beside these signals will get yet much faster and to farther distance but under this space field and below sub atomic particles. These gravitational, magnetic and other forces in cosmic space are responsible for passing signals in all space and between objects which are not always noticable.

When will be made vacuum even below atoms if possible then the same in other place and even on other planet and from this vacuum will be send signals to other place should go not that noticable for space here, yet when from such matter in vacuum will be make waves to send signals to other similar device then should pass faster than light and will be not notice here also.

So this include sending information by waves and radiation all over space.

This information can be eventually recover from atomic and subatomic particles which is coded there like on much smaller scale memories information are in atomic and subatomic particles in brain and sometime subconciously come out from there under some memories from past and one may start sing out of nothing some song which was hear time ago, or recall some details from past which are decoded in small atomic particles in brain and yet very small part get out from all these past experience, But these elementary in brain particles are only very small part all this space sphere in micro world atomic and smaller particles. In brain there is whole connected unit of atomic particles so can create whole world of information but from one single atomic particle everywhere will come single information which was send and receive.

Not all these send signals will get anywhere becouse network connection sometime dont work as should and some forces may stop it but yet in smaller particles than neutrino they do it in cosmic space or even by deeper parts.

To apply in such magnetic (or other) field or some object or system- to analise reaction with for instance 10 millions elements or atoms should get in some case five millions two different elements each. But could happen that in scale of whole field one seperate elemnent or part one seperate element of it have five millions sub elements and othher four millions eight hundred thousnnd becouse rest on scale of whole field could be of attracted by other elements or may in some case pass out of whole field. So in particular area of field or rest will be left nine millions eight hundred thousand two seperate elements. So by this law of constant permanent relations can be apply to whole field not to particular area of field. In case if some get out of whole field then cant be apply to whole field.

Unlike as in chemical reactions in atomic reactions is considered that are not meet fraction reactions. Though it depend from way of look on it.

But many atoms have also many electrons around so during reactions or divisions particles some electrons could be lost or attracted to other particles.

So weight realations could be 0,9 for instance to each particles or their groups, and 1.1 to other Also have influence condition of place where reaction happen like cold or hot yet other.

Permanent action is directly proportional to value which create it. So value of whole system is directly proportional to action all elements it contain.

Though other system from outside also may have influence but this also depend from value first system. For example is there hot place or room then oustide cold have some influence to lover temperature inside. But then if temperature inside will be increase more then can have more influence outside.

Or if from space some elements or forces come to earth also make some change on planet. Yet if electromagnetic forces on earth will increase or more denset atmosphere then may lower outside influence or reject it.

Between loose atoms of hydrogen for instance in plenty space in comparison so there most be some chunks run between atoms as in space are chunks all kind between stars and planets. Chunks there sometime attach themself to atoms and this make some difference even slight on atoms forces or weight but they are to small to be detect by known technology.

As atoms by motion create thermal energy so by use hightly sensible and on micro scale detecting devices is chance by this to detect thermal energy coming from under waves, yet to find which way atoms move or spin around as with spin around is more greater thermal energy which may show up.

Important from all of this find new ways and new theories dealing with forces of nature and it matter not just stick to these ways and theories which are known now. For example theory of Big Bang which is unnecessary so popular now but should never be consider in first place between mostly scientists if they are so really developed. As this theory is without any sense and against logic. If there been Big Bang explosion so there is quite obvious thing that to have explosion is need to have space. Bomb or other object will not explode if there is no space, denset or thin. This against law of physics and common sense also. But many still have believe in these kind of theory.

When itself show picture that first must have been space there anyway; that is three dimensional but without any stars or other objects known now which start come later from below this known field.

Similar need to be use new ways and theories in other field of science and also in millitary strategy where are still some old in use strategies and ways of actions, instead find new ones which may bring better result in action than these known so far. There was before occasionally ways and strategy which bring victory when was reject as rather impossible by others. So if want to win then better think of new ways too and these one also which other will think would be impossible to do it.

There was in past millitary activity which was unnecessary ignored like for example rather impossible to cover some terraain to attack form such side but it happen and bring victory. Yet similar examples happen also. So always better be prepare for even most improbable possibility and try them sometime too to have better result.

Symetrical arrangement atoms also could change structure or shape particles.

Have to be take also under considerattion forces activity between atoms which may too change somehow during this arrangement.

Of course not only radioactive material show radiation but also other kind of body.

Just their radiaction is not that strong and visible in micro scale or macro.

The same is rather impossible to predict which atoms will devide by some period of time and have to be statistic measure especially if their number is very large.

But when is known reason for such division prediction may become more reliable.

Like outside influence, radiation from outside factors, and from which side is stronger. Or location and set atoms in particular body which is never identical so this part with smaller amount may divade faster if there is stronger outside influence. Also structure of atoms is never the same as atoms alone differ from each other, one with stronger feature than other. Beside other conditions.

But there are ways to analise them too. If some comet or meteor come close to the earth nobody could known it. Yet if telescope will be use it maybe notice and from mathematical ways predict more direct path it will pass.

Here on micro scale by observing by longer period and many time division in structure radioactive or other body from every side and in micro scale too wll be possible eventually predict with more probability division particular atoms of the body. Or yet in computer program watch division and trace progress of division. With going also some changes during such process.

Help would be too if is known divsion one or few atoms in process then count their even approximet numbers in body so would be more or less known whole their division particular place in body and few other factors.

Large part with great numbers may show more relable statistic prediction than smaller numbers but if divade to few sections with smaller amount statistic would be not so accurate. But with repeated experiment few time it may show better result as to each or most sections.

But some simple way may help come close to them. Not only search subject have to be place in vacuum and inside part of microscope as well but yet whole chamber where search is done better make vacuum as much as possible. Search subject could be put on big screen to have better view and try to increase as possible without losing clear picture. Still atoms will be under light waves as small pebbles are under water not vissible. But these small pebbles emit some waves which can be pick up by sonar. Similar here but atoms when in motion have radiate more waves which come to the top and leave some traces which maybe pick up when bombarding atoms with alfa rays for example. Sometime they get to the atoms nucleus which will radiate more and often to top of light waves which cover them and there leave trace. If will be stronger repulsion from nucleus then more chance get to the top.

Two atoms when push to each other with similar sign will repulse each other. If light waves which cover atoms will be very little magnetise so with now stronger magnetisation they could be pull up or spread on sides as water which cover pebbles will be magnetise later with stronge electromagnes and try pull up or to spread.

If in small chamber will be done as much as possible vacuum and will be slight magnetise so walls from all sides will be turn as very strong electromagnetic field to pull outside what is left inside to receive yet bigger vacuum and spread all particles. Then may repeat it few time. Then also can be use microscope for search remaining field with elements there which been still left.

Atoms with similar sign will repulse each other but when will be apply very strong magnetic field they will be pull to each other yet with bigger result with application very fast rotating movement.

Have to be consider again that atoms beside nucleus, neutron and electron consist yet other particles like in solar system beside sun, planets are also moons and yet smaller objects inluding asteroids etc. Similar in atoms also may run or spin around other particle or particles but they are to small with little mass to be notice or detected.

All atoms becouse of speed movement create around energy field which are almost exactly the same as are atoms and they still are for long time after atoms destruction and by this create something like next similar world after this one and can easily pass this physical one here as they are not atoms.

All this going similar way up to macro world as well that is repeat similar process.

Beside have to be mention yet smaller particles between atoms which was result during atoms creation and their movement. In Wilson chamber even with more vacuum maybe chance to observe it if they left any trace during radiation activity or have any slight influence on beta rays if are close to it as electron have very small mass. If more vacuum in chamber could have better result as is not so dense matter- when is more denset matter is smaller chance to see any trace from these small by size and mass particles. Also other smaller yet particles which was of course result during atoms creation and without doubt are between nucleus and electrons and between atoms which when collision between them happen they lost some of mass part in shape small particles and it happen too when electrons are lost from atoms. Occasionally though less is show on great scale when planets or moons have collision and plenty of particles are left later. This have small infuence on atom movements.\ in micro scale.

If some particle dont have electric charge it dont mean that cant be bent as maybe apply for instance more denset light beam, temperature or else.

And too could be magnetise.

As all this come from atoms movement which are also made from yet many smaller particles including nucleus or electrons so by this movement they send number of particles, radiation in shape of quantum or others.

With slower movement of throw quantums it will show light as quant particles. But if will be still faster and faster speed of throw the quantums there will be decrease space between them until they will show up as a waves not quantums. In this micro world could be seen still difference in space between quantums but here to obserwer will show waves character.of light.

So speed alone change character of light. If particles will be throw with speed one per second it could be seen quantum character but if will be throw with speed many thousand or millions yet much more per second one after other it could be seen here as waves character. Do not mention again that atoms send radiation and elements other than these known but not discover yet.

Still deeper in vacuum there is way to discover it.

This all depend also from other conditions like temperature and spin around faster and faster quantums. So then in dependence from movements and conditions it can be more numbers than just duality of light waves and so this could be apply to other objects or particles.

Have to be done in chamber much more vacuum space yet fill up with lighter and less denset substance. When in accelerator in the end in bigger vacuum atoms are destroy and next pass to this chamber will be better chance to find out these smaller particles which have different radiation than whole atoms but not detected becouse of small size though they are together in whole atoms radiation but dont notice if all together.

If reach below atoms in laboratory it show only kind of dark matter becouse not that much radiation as atoms created, but is possible by whirl pull up to small micro needle like pipe to make this matter yet more denset as will accumulate more in smaller space , next search for it characteristics and next then spread out in gravitational field to search also for particulars and characteristics such matter. This matter under atoms and sub atomic particles is some like polarisation field from coming there yet under very sparse particles which when come to this field by speed show light like points as result of this speed movement and matter there. Is also possiblity to attract this below atoms more dark matter by strong gravitational field to hold and search then or first radiate it with magnetic field and next attract with yet stronger magnetic field. But in this kind in Wilson chamber have be create very thin matter if want to see trace of coming out these very small below atoms particles.

Isotopes are consider the same as other elements with the

same properties just different mass. But this is rather not whole true becouse if all properties will be the same so should be mass too which mean that must be yet some different property or properties which make different mass; like different moton in atoms and sub atomic elements as different speed motion when especially rotate can create different mass or yet some other difference there and outside influence. As was mention is different radiation coming out from elements which collide and different when spin around.

Of course usually hot pass to cold under known here condition but if hot object or matter will have much stronger gravitation or magnetic force then cold will pass to hot object or matter not reverse.

Black hole pull in objects or matter whatever they are cold or hot.

Uncertainty theory is now popular becouse was not find ways for improvement of it. But all depend too from ways looking on subject and finding new ways.

If there is some town with main street and people walking on sides, cars run in many direction,, then when travel there would be rather not known what will happen ahead yet in farther distance. Which people could be meet or pass ahead, which cars will come from other direction or side street to main road.

But if one will be looking on town from high hill then will see people and cars moving in some directions. So from hill will be known or more predict what other people will meet or pass ahead, or what passengers in cars will pass in farther distance, or coming people and cars from sides to main street yet from main street. So maybe even some incident if speed is to fast.

Cars will be run with fast or very fast speed so would be difficult to predict in which point will be farther on street yet with some traffic around or street works maybe ahead. From top of the hill will be more easy to calculate in which point on the street cars will be when can see what is ahead.

So uncertainty gradually dimish to observer from hill

though not for people in town.

Similar if some object or just coin will be turn very fast around so will be rather unknown where in particular point choosen side of object, coin will be in 360 degree around, yet will be not seen particular side of objet or coin.

But if this will be show on screen then by slower and slower movement done on screen will be possible to see all details necessary when was not available to observer out of screen. However even beside outside screen if is known mass of object, coin and is measure and known speed rotating object around and distance to move around 360 degree so is possible to known where in particular point each side of object, coin will be and in which side around.

The same with some moving object in distance ahead. Also in this case maybe use prediction or calculation . If in Universe space or some room will be plenty of objetcs, elements- spin or move around or ahead then when looking from outside and is known their numbers would be make good calculation of their ways and path, more exact movement and end of the result.

So from outside when look is more easy to predict in which point will be some element or elements yet when is known their speed and movement other elements around and forces acting there. But for particular object, element in Universe or room this will be not possible becouse for it is not known what is ahead and if some other elements or forces will not stand on the way when move. Even if not move around still will be not known for element, object if other elements or force will be pass to attract it or move to other place, or yet other not known to this element functions.

But next if will start to move away from Universe or room and from farther distance will be see many of them then chance to calculate or known these measures before ways, paths become more difficult or impossible not to mention distance between them when moving away.

If some vehicle on street or object in space is moving with some speed even turn around so when is known lenght of distance then is more easy to calculate in which particular point it will be. If speed change so have to be take under consideration particular section of distance it cover. So if is known length and speed then can even predict when in given moment object will be in some point, distance.

Similar on atomic level if is known mass and speed atom or electron and a distance on leght or spin could be calculate where it be in given moment or number of them not just one, yet when observe from distance.

So uncertainty is rather to be bring to particular area not whole system.

Also is good to find ways when reach for some problem to solve. These ways and predictability maybe use in life, science activity or in defence.

If in some area, room are known round objecs about meter across and many of them is possible to calculate when they spin slow in which point market sign will be or if move on some distance also. Then if speed will become much faster will be also possible to known in which point market sign will be yet when is known speed and distance. If objects will be 10 time smaller about 10 cm. so also should be no problem with calculation by divade it by 10 and known speed and distance yet forces around. Next when come to atomic level also next have be calculate how much atoms are smaller than last describe before objects and distance they cover with known speed and forces around. So may get enough close to find proper measurements closing to certainty in this case.

Not always is need to known all details and elements becouse when search for some result so just few known details, elements or forces would be sufficient if desire result is achieve.

Help will be too if will be known each movement of forces like gravitation or other which are in all space from smallest to bigger objects. For example could be calculate speed and force and distance where objects will be either big or very small when earth gravitation will catch particular object to bring here or spin yet first around. These forces are also in micro world just speed there is different. However speed is only increasing force or mass of object so to know it increasing force or mass would be calculate speed too, though have to be consider also mass and forces around . Yet to known where some object will be in given moment. If in large room will be move, spin bigger, small and very small objects many with different speed so from outside will be possible detect general energy and vibrations in this room. But becouse objects have different mass and also speed so will be possible to find in which part of room is more energy, gravitation or other forces. Yet with proper equipment up to micro movement. Now if these objets will change position or speed so could be next possible to find out by change these forces in which place they are next and if they change speed without going inside. Whole room have sum from all these objects in general but in each part of room will be there much variations so by these variations and change of forces will be chance to find change speed and position.

In atomic world if will be apply magnetic field or stronger yet if need this will slow down movement particles maybe enough in some case to observe desire moveent and repeat observation few time to receive proper data.

So later will be known or maybe more exactly calculate their movement, whole speed and position.

Chaos theory also maybe take under few ways when look on this.

One thing is that small cause can bring big or very big effect or change or in reverse. When just change one letter in one word it could change to different meaning or even for whole sentence. Sometime just dot in word would change meaning and also in whole sentnce.

In big construction if only one screw would be wrong done or to weak may happen that this big contruction coudl be destroy. Yet opposite often very big effort could bring very little effect.

If order would be mark as 1 and chaos as 1a then 1b which will be even bigger chaos. So when 1 will be destroy become 1a that is chaos. And elements there will be consider as chaos. But if 1b would improve and get then to 1a so then 1a will be consider for 1b as order when the same elements for 1 will be chaos. So this depend from conditions and elements view.

When plants die will be consider chaos but they later improve soil and other plants will grow better so chaos improve order that is 1. Beside number 1 which present chaos could be number 2 in this case for instance represent homogenous matter which from this point here maybe consider different occurence. If then number 2 will mix with number one would improve it or degradate to number 1a or number 1b independence from occurernce 2.

If will be mix with 1a or 1b it may bring it to number 1 or farther degradate it. Beside number 2 maybe next other numbers describe known or not known occurences when will be discover later.

As again to the colours bodies show different ones becouse some waves was absorb but some rejected. Though is only part of process. Bodies are made from atoms which becouse of movement radiate waves also so this radiation have own colours too. Is depend from what kind different atoms body is made, their amount, movement etc. If body will not absorb waves just have own only waves radiation colours will be different but as they absorb some waves these waves with connection with atoms waves create yet different colours than just atom have itself. This of course apply to others process not only colours radiation. If white light is pass by prism this prism made from some material have own atoms to so by mix with white light show together differnt colours. If prism will be made yet with little different material then will show slight difference in colours than previous one.

Electromagnetic field save earth also from radiation but to

some degree maybe done increase of field if in need. In general could be set in north pole and south as well, high over ground plenty magnets or electromagnets then turn them around with speed similar to earth rotation so with time would made around more electromagnetic field stronger or weak in dependence from amount and force applicated devices.

In case of these kind of war as they may happen electricity and modern technology could be devasted or by some natural catastrophy.

Have to be back then for while to old methods and technology at least in some areas. Like back again to use steam machines which with farther development may bring some help. With majors steam engines working if want to transport something farther away can be build on the way smaller size steam engines to transport even for long distance. Beside some cars could run on steam engine as was done before.

To use just computers as only source information is not best way becouse if electricity gone much could be lost so is always better to use beside them the same information by hand writing or make some copies on paper as discs and similar copies will be useless too.

In case of building structure have then to be back to old methods use levers which was probably also in use to build piramids or other acient structures becouse with strong solid base very large objects can be lift up or pull by many hands or with animal help.

Crops if necessary will be possible plant in green houses worm up if need by working steam engine. Beside small generators maybe possible to use in homes or other buildings if wires was not destroy or then change for new one.

If not other way generators maybe rotate by hand using wheel for while to produce electricity or change batteries.

Have to be make sure that will be eneough water available and not contaminated.

Amount of water from at least millions of years is about the same circulate to atmosphere and back. When in start of 20 century was about one billion people on earth and now few time more so use now about few time more water which eventually get to smaller amount, plus much more domestic animals which also use water and much bigger industry to use plenty of it.

May not come to such case but better to known what to do and even now some could use these methods for own purpose.

2022

www.ingramcontent.com/pod-product-compliance
Lightning Source LLC
Chambersburg PA
CBHW040518220526
45473CB00012B/2901